Contents

Foreword:.. 2
Hydroponic Fodder.. 3
Benefits of Hydroponic Fodder... 5
Disadvantages of Hydroponic Fodder.. 6
Composition and Nutritional Analysis of Hydroponic Fodder...................... 8
Hydroponic Fodder Analysis..10
Differences Between Traditional Forage and Hydroponic Fodder..............11
Where can Hydroponic Fodder be Produced?..13
Hydroponic Fodder Production Steps..18
Factors Influencing Production..33
Systems and Shelving..41
How to Build a Homemade System...45
Irrigation System...46
Drip Irrigation..49
Horizontal System:...50
Nutrient Solution...52
Preparation of the Nutrient Solution...55
pH (hydrogen potential) & EC (electrical conductivity)................................63
Application of the Nutrient Solution..67
Steps to Calculate Hydroponic Fodder Irrigation Needs.............................70
Use of Hydroponic Fodder for Different Species of Animals......................74
Bibliography..79

Foreword:

It is my pleasure to present this book on hydroponic fodder production, a work born out of my deep passion for agriculture and animal welfare. Over the years, I have had the privilege of acquiring knowledge and experience in this field, and my aim in writing this book is to share all that learning with you, dear readers.

The primary reason that drove me to write this book was the belief that every household, especially those with farm animals, should have access to valuable information about hydroponic fodder production. In an uncertain future, where the scarcity of traditional fodder could become a reality due to droughts or extreme climate changes, this knowledge could be very useful in ensuring proper nutrition for our animals.

Within the pages of this book, you will find all the necessary information to successfully produce hydroponic fodder for your animals. From basic concepts to advanced techniques, each chapter is designed to be a practical and comprehensive guide that enables you to carry out this process efficiently and effectively.

Furthermore, I want to invite all readers to follow me on social media, especially on YouTube, where I share additional content in the form of instructional videos. These videos demonstrate, step by step, how to produce hydroponic fodder in a simple and accessible manner, with clear explanations that everyone can understand.

I hope this book serves as a valuable tool in your journey towards more sustainable agricultural production and responsible animal care. I wish you much success in your hydroponic adventures!

J.E Ornelas

You can find me at: YouTube: Jesus Ornelas
https://www.youtube.com/channel/UCPZvK9GT0swn6sauo8w0hDw

Hydroponic Fodder

This is like preparing fresh food for animals using a special technology called Hydroponic Fodder. It's like a food factory where germinated grains such as corn, wheat, oats, and barley are used to feed animals in controlled environments. hydroponic fodder is highly nutritious and of high quality. It develops rapidly and can be harvested within a minimum of one week and a maximum of two weeks, providing a rich source of proteins, minerals, and vitamins for animals.

The great thing about this feed is that it replaces traditional fodder without affecting animal nutrition. It is calorie-dense, with easily digestible fats and carbohydrates. Additionally, it does not interfere with the animals' metabolism.

This type of fodder can be cultivated year-round and anywhere, even without the need for soil! It uses the hydroponic technique, which allows plant cultivation without soil but requires specific conditions.

In hydroponics, various plants are cultivated for optimal growth, taking advantage of the ability to produce more plants than traditional methods due to controlled growing conditions. This method focuses on supplying water and nutrients directly to the plant roots, without the need for soil. In the case of hydroponic fodder, the same technique is used, but the plants do not reach the fruit production stage; they simply germinate, so their nutrient needs are minimal, sometimes only using potable water to grow.

Although "hydroponics" may give the impression of working with water, in reality, this form of cultivation helps conserve water, which is excellent. It's an ideal option, especially in regions affected by droughts or harsh winters. A "hydroponic chamber" is created to protect the plants from the sun, rain, and cold, where perfect conditions are created for optimal growth.

The first step is to select the type of grain, such as wheat, oats, barley, corn, or rye. These grains meet the necessary requirements for optimal production.

The grains are sown and germinated daily in soilless trays placed on shelves. The trays are watered simultaneously to ensure adequate water supply from top to bottom.

hydroponic fodder is an excellent option for feeding a variety of animals such as lambs, pigs, goats, calves, dairy cows, horses, rabbits, chickens, ducks, guinea pigs, turkeys, and other farm animals. It's a very effective way to provide food for animals!

Benefits of Hydroponic Fodder

1. High Nutritional Value: This food, known as hydroponic fodder, is excellent for animals because it contains many healthy components such as proteins, enzymes, carbohydrates, vitamins, and more. It helps animals stay strong and prevents them from getting sick.

2. Effective Use of Space: The way they organize trays on shelves in thermal chambers efficiently utilizes space. With a setup of 240 trays, they can produce 740 Lbs of fodder per day, occupying only 75.6 m2. This is much less space than required for traditional farming, almost like having one and a half hectares of land!

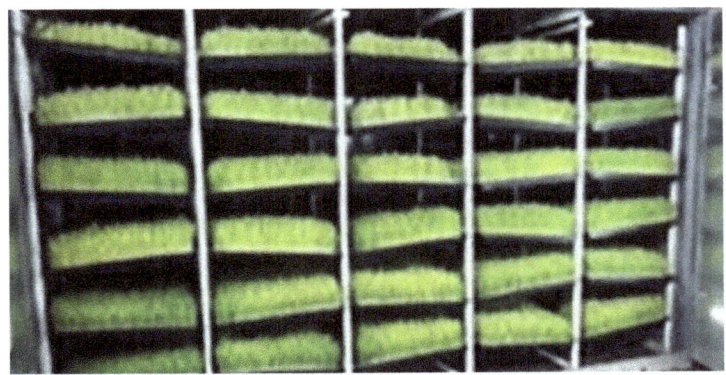

3. Water Savings and Efficient Use: hydroponic fodder saves water compared to traditional systems. The amount of water required to produce hydroponic fodder can vary depending on various factors, such as the type of grain used, environmental conditions, and cultivation methods. However, generally, it is estimated that around 0.5 to 1.0 gallons of water are needed per 2 lbs of hydroponic fodder produced, whereas to produce 2 lbs of alfalfa or corn in soil, 40 to 80 gallons are needed. This is because in hydroponic fodder, water losses are very low.

4. Short Production Cycle: The time to produce hydroponic fodder is short. Up to 3 harvests can be obtained in 30 days, which adequately covers the animals' food needs. This is great for keeping animals well-fed, especially for dairy cows and cattle.

5. Low Operating Costs: Compared to traditional farming, hydroponic fodder is less risky in the face of climate change, uses less land and water, and has a lower risk of crop diseases. It does not require machinery, and labor costs are lower. Additionally, hydroponic fodder brings more benefits such as increasing milk production in cows, improving meat quality, increasing the weight of young animals, and many more. It is a good economic option for farmers and helps them have healthier and more productive animals.

Disadvantages of Hydroponic Fodder

1. Misinformation and Overestimation: Sometimes, hydroponic fodder projects are marketed as a perfect solution to farmers without truly understanding their needs. This leads to failures because producers are not familiar with system requirements, such as plant needs, pests, diseases, and ideal conditions of light, temperature, and humidity. Lack of prior training can result in issues in system management.

2. Commitment and Continuous Care: hydroponic fodder requires constant care and a solid commitment from the producer. Without the correct information and necessary knowledge, managing the system properly can be challenging. The process is continuous and demands constant attention.

3. Lack of Knowledge and Information: The absence of simple and direct knowledge can become a disadvantage. Just like with familiar hydroponic technology, a lack of information can lead to problems. It is essential to have knowledge about system management to avoid difficulties and ensure the project's success.

4. Low Dry Matter Content: Hydroponic fodder has less dry matter compared to other types of fodder. Therefore, it should not be the sole source of food. Each type of animal requires different foods, and hydroponic fodder is just one part. It is crucial to use other feeds, such as straw, to complement the lacking dry matter in the animals' diet.

These disadvantages highlight the importance of training and proper knowledge for farmers wishing to implement hydroponic fodder systems. Without this information, projects may face difficulties and fail to achieve expected results, including the low dry matter content that can affect animal nutrition.

Composition and Nutritional Analysis of Hydroponic Fodder

The quality of hydroponic fodder can vary for various reasons. It is crucial to highlight this point, as often this information is not adequately disclosed to farmers. Some companies selling production equipment may mention that hydroponic fodder provides more than 18% protein. However, this information is incomplete if the necessary requirements to achieve these protein levels are not specified.

1. Harvesting Time: Harvesting time is a crucial factor that affects the quality of hydroponic fodder. Depending on when it is harvested, this fodder may contain more or less protein and fiber. For example, if harvested early, it is likely to have more protein, but if left longer, it is likely to have more fiber. Regardless of the grain used, all varieties of hydroponic fodder have a period when they reach their **optimal nutritional values**. It is during this period that harvesting is recommended to ensure the best nutrition for our animals. Outside of this period, nutritional levels may be lower.

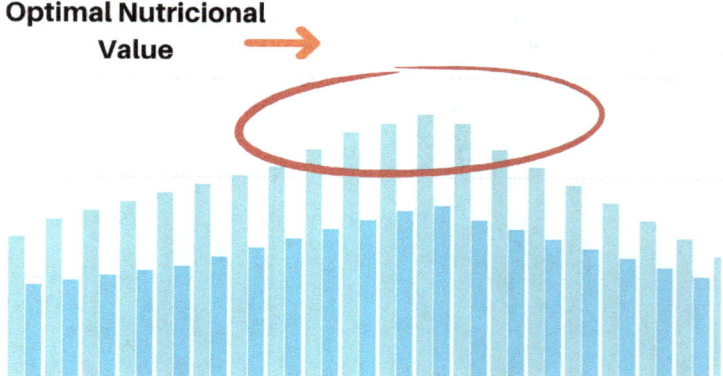

2. Plant Age: As plants grow, their composition changes. Young plants tend to have more proteins and less fiber, while older plants may have less protein and more fiber.

3. Type and Variety of Seed: Different seeds contain different nutrients. Some seeds are naturally richer in certain nutrients than others.

4. Climate: Such as temperature, humidity, light, and water, can affect how plants grow and what they contain. Favorable climate can lead to plants growing more and being more nutritious.

5. Crop Management: How plants are cared for also matters. Using fertilizers, watering them properly, harvesting when necessary, and protecting them from pests can make hydroponic fodder better for animals.

So, the quality of fodder can change due to many factors. Understanding this is important to ensure that animals have good quality food and that plants grow well.

Hydroponic Fodder Analysis

Next, we are going to compare two types of hydroponic fodder. Both use corn seeds, but the difference lies in how much time passed before being harvested. The first one is harvested in 11 days, and the second one is harvested in 14 days. There is an important aspect to note: as more time passes, the protein level starts to decrease, and the dry matter level increases. In other words, the feed has less protein and more dry material as more time elapses before harvesting.

Nutritional Analysis - 11 days

Dry material	18.60%		Calcium (Ca)	0.10%
Protein	18.80%		Phosphorus (p)	0.47%
Metabolizable Energy	3,216 kcal/kg. MS		Magnesium (Mn)	0.14%
Digestibility	81-90%		Iron (Fe)	200 ppm
Carotene	25.1 UL/Kg		Manganese (Mg)	300 ppm
			Zinc (Z)	34.0 ppm
Vitamin E	26.3 UL/Kg		Copper (Cu)	8.0 ppm
Vitamin C	4.5 mg/Kg			

FUENTE: FAO

Nutritional Analysis - 14 days

Dry material	+22%		Calcium (Ca)	+0.18%
Protein	-16%		Phosphorus (p)	-0.34%
Metabolizable Energy	-2,600 kcal/kg. MS		Magnesium (Mn)	-0.26%
Digestibility	-64-68%		Iron (Fe)	-79 ppm
Carotene	N/A		Manganese (Mg)	N/A
			Zinc (Z)	+48.0 ppm
Vitamin E	N/A		Copper (Cu)	+15.0 ppm
Vitamin C	N/A			

FUENTE: PCTI

Differences Between Traditional Forage and Hydroponic Fodder

1. How it's Cultivated:
 - Traditional Forage: Planted in soil using regular methods.
 - Hydroponic Fodder: Grown without soil, in water, using the method called hydroponics.

2. Required Space:
 - Traditional Forage: Requires more land space for growth.
 - Hydroponic Fodder: Utilizes shelves for vertical growth, saving space.

3. Land Usage:
 - Traditional Forage: Relies on soil for nutrients.
 - Hydroponic Fodder: Doesn't require soil; nutrients can be directly incorporated into the growing water, achieving optimal results even using only water, without the need for additional nutrients.

4. Water Needs:
 - Traditional Forage: May require a lot of water, especially in drought conditions.
 - Hydroponic Fodder: More water-efficient.

5. Time to Grow:
   - Traditional Forage: Growth time can vary depending on the type and climate.
   - Hydroponic Fodder: Grows faster, typically ready for harvest in 7-14 days.

These differences highlight why Hydroponic Fodder can be a more advantageous option in certain situations compared to traditional forage.

Where can Hydroponic Fodder be Produced?

There are several places where we can install vertical systems as seen in these images. The most important thing is to ensure that we provide the appropriate conditions for growth and development. Let's remember that the better we control these conditions, the better our results will be. These conditions are divided into four parts, which we will detail in the next chapter of this book: Temperature, Humidity, Light, and Aeration. Improving these four factors to optimal levels will give us better results.

These systems are adaptable both in outdoor environments and indoors, such as rooms or warehouses that are modified to provide the appropriate conditions for hydroponic fodder cultivation.

Climate-Controlled Container: These systems are the ideal option for hydroponic fodder production, as they offer maximum conversion in terms of final weight and nutritional value. They can transform 1 lbs of grain into up to 11 lbs of hydroponic fodder.

Designed with specialized modules, these systems ensure optimal control of the four fundamental factors for production: Temperature, Humidity, Light, and Aeration. Although the initial investment may be significant, the advantage gained is the ability to produce hydroponic fodder every day of the year, regardless of weather conditions. This approach offers a long-term solution that can be profitable and sustainable for animal feed production.

Indoor or Closed Place:

Similar production to that of a climate-controlled container can be achieved as long as control of the four factors is maintained: Temperature, Humidity, Light, and Aeration. It is important to remember that when these factors are not controlled, growth and development can be slower.

As seen in the images, the first one shows a production of 10-11 lbs of hydroponic fodder per lb of seed, thanks to the use of artificial light and 100% control of the four factors. In comparison, the second image shows a production of between 6-8 lbs of hydroponic fodder per lbs of seed, as artificial light is not used in this instance and the four factors are not optimally controlled. These places often require a special focus on light and aeration, ensuring adequate airflow and minimal lighting. Artificial lighting is common in these facilities to maintain optimal growth conditions.

Greenhouses: This system allows for optimal growth and development for approximately 12 hours a day, as during the night, the plant experiences metabolic slowdown due to decreasing temperature and the absence of light.

When considering the installation of a greenhouse, it is crucial to ensure that the temperature does not exceed 86°F, as high temperatures can cause stress on hydroponic fodder and promote the development of fungi and bacteria.

This method is commonly applied in regions with cold climates, highlighting the importance of temperature control to achieve optimal growth and development of hydroponic fodder.

It is crucial to note that the final results in the production of hydroponic fodder can vary significantly depending on the control of the 4 essential factors. Yields can range from 6-8 lbs of hydroponic fodder per lbs of seed to reaching 10-11 lbs of hydroponic fodder per lbs of seed, all depending on the care and precise control of these 4 critical factors: temperature, humidity, light, and aeration.

Outdoors or Open Spaces:

This type of setup is suitable for various outdoor spaces, with the primary condition being to avoid direct sunlight, as hydroponic fodder does not require intense light exposure. It is crucial to ensure that the temperature does not exceed 86°F, making this option less suitable for warm or humid climates unless temperature control can be adequately maintained.

This setup in open spaces may face challenges related to rodents, squirrels, birds, and insects. Therefore, it is primarily recommended for small-scale productions or initial trials.

It is important to note that in this type of configuration, an average yield of 6-8 lbs of hydroponic fodder per lbs of seed can be expected. This is because the 4 critical factors (temperature, humidity, light, and aeration) cannot be controlled 100% during the 24 hours of the day, with growth being slower during the night when light and temperature decrease.

Hydroponic Fodder Production Steps

Now, we will explain the steps for hydroponic fodder production. These steps apply to both small-scale and larger-scale productions. In this section, I will guide you through the necessary steps to cultivate this type of hydroponic fodder using primarily corn, barley, wheat, and oats seeds. These steps are simple and effective, providing a clear overview of the production process.

-Production Steps:

1- Grain selection
2-Wash and clean grains:
3-Disinfection:
4-Soaking grains:
5-Sowing:
6-First phase/darkness:
7-Second phase/illumination:
8-Harvest:

1- Grain Selection:

You can use various types of grains; it's not necessary to have special grains or seeds. The most important thing is to choose grains that can germinate. Remember that any grain that doesn't germinate will be more prone to the development of fungi and bacteria.

The grains most commonly used are corn, barley, wheat, and oats. However, other alternatives such as rice, lentils, chickpeas, among others, can also be explored. It is crucial to note that sorghum is not widely used in production because when it sprouts (or re-sprouts) and is tender, it has the peculiar ability to produce toxic acids, such as cyanide acid. This mechanism is the plant's natural defense against insects. If you consider using sorghum, it is imperative to conduct analyses to determine the presence of this toxin.

Selecting grains with a high germination rate is essential to ensure success in your crop. It is advisable to opt for grains harvested in the last 12 months, as their germination capacity tends to decrease with storage time. However, this does not mean that grains stored for 2-3 years or more will not work; it simply implies that the more grains germinate, the better the final results will be.

Make sure the grains are in good condition, without damage, breakage, insect attacks, or presence of spores or rot. In the following image, we notice grains with small holes, indicating that some insects have already started feeding on them. These grains probably will not germinate, and if they do, they are more likely to rot, which could cause fungus and bacterial issues. Additionally, grains with a darker color are observed, indicating the presence of bacterial spores and the beginning of the decomposition process.

18

In this other image, upon closer inspection of two grains, we notice that the grain in the background did not germinate because insect damage led to the death of the grain. In contrast, the second grain shows signs of sprouting despite the damage. However, this second grain has already started to decompose, which will result in the development of fungus and bacteria over time. This decomposition can negatively affect other nearby seeds or grains.

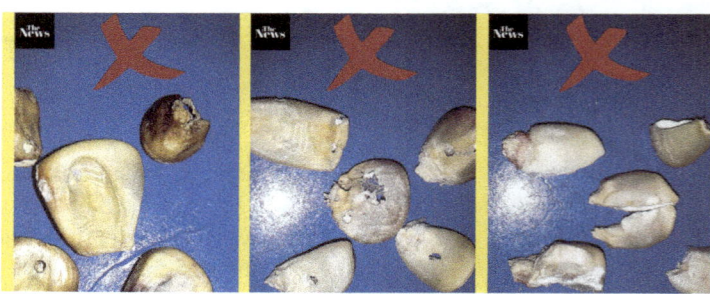

When selecting grains, it's important to note that there shouldn't be too many grains with imperfections, as shown in this image. While achieving total perfection is difficult, the goal is to minimize the number of grains that will not germinate.

Avoid using grains treated with pesticides, as they are intended for outdoor planting and contain chemicals that can be harmful to our animals. These grains treated with pesticides are easily identifiable, as they are usually completely covered by these chemical products. For example, pesticides often have a green or red color.

Preferably, use local grains to support the economy of the region and, in general, they are usually more economically accessible than imported ones.

In this step, the main goal is to achieve germination of as many grains as possible. It's important to note that achieving total germination is difficult, but the aim is to maximize this process for optimal results. To achieve this, it is recommended to select grains with a germination rate ranging from 90% to 95%. This will significantly contribute to the success of your crop.

2-Wash and clean grains:

In this second stage, the goal is to remove all impurities present in the grains, such as soil or small particles. However, the most crucial aspect is to remove the grains that will not germinate, including those that are broken, damaged by insects, or infertile. The easiest and most economical way to carry out this process is by using water, as the grains that will not germinate tend to float on the surface.

In this step, the process is quite simple. Clean water is added to the container where the grains are placed. All impurities will float on the surface, and we will use a strainer to remove them.

Note: It's important to stir the water with the grains to ensure all impurities are removed. The amount of impurities depends on how the grain was processed. If modern machinery was used, there would be fewer impurities compared to traditional processing.

21

3-Disinfection:

This step is essential to ensure the elimination of any bacterial spores that may be present in our grains. These spores are nearly invisible to the naked eye and can be difficult to detect, so it is necessary to ensure their removal. Their presence without elimination can promote the development of fungi and bacteria in our hydroponic fodder.

In this stage, we will use clean water and common household bleach (sodium hypochlorite), allowing the grains to soak in this solution for 15-30 minutes.

The amount of chlorine may vary depending on the percentage of sodium hypochlorite, as products on the market have concentrations ranging from 4% to 6%. The most commonly used dose is 2-4 milliliters of chlorine per liter of water (0.06-0.13 fl. oz.) . It is crucial not to exceed the soaking time beyond 30 minutes, as prolonged exposure could negatively impact the grains.

Note: It is also possible to use calcium hydroxide, known as Agricultural Lime. The most commonly used dose is 5 grams per gallon of water.

4-Soaking grains:

In this step, we initiate the pre-germination process to activate the plant's metabolism and transform the grain into a seedling. The goal is to hydrate the grains so they absorb all the water necessary for germination.

To carry out this process, we simply soak the grains in clean water. It's crucial to ensure that the chlorine water has been removed and to use fresh, clean water. Corn generally requires around 24 hours of soaking, while smaller grains like oats, wheat, and barley only need about 12 hours.

In warm climates, it's recommended to oxygenate the grains halfway through the recommended time, meaning to oxygenate the corn at 12 hours and the other grains at 8 hours. Oxygenation simply involves removing the water and leaving the grains without water for 1 hour to allow oxygen to enter. Remember that grains are living organisms and need to breathe.

Another way to determine if the grains need oxygenation is by observing bubbles in the water, which indicates that the grains are releasing gases and need to breathe. It's crucial to note that the presence of bubbles in the water is completely normal. However, when there is an excess of bubbles, it indicates that the grains are releasing a significant amount of gases and, therefore, need to be oxygenated.

Note: It's important to add enough water, as the grains will increase in size by 15% to 35%, depending on the variety.

5-Sowing:

Once our grains have been washed, disinfected, and soaked, the next step is to place them in the location where they will germinate. It is recommended to use trays specially designed for hydroponic fodder, but other alternatives can also be considered, as seen in the second image.

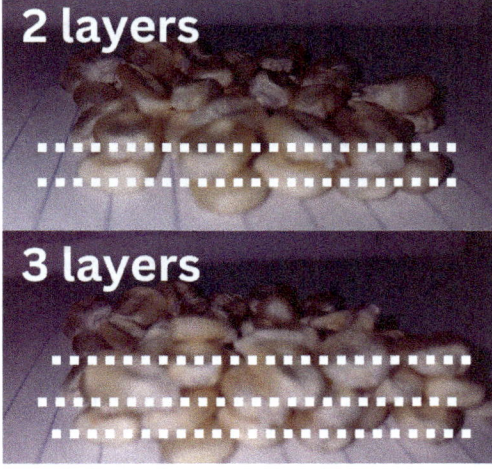

The amount of grains can vary depending on the type and variety of grain, as well as their size and weight. The most commonly used amount is 2 to 3 layers of grains, avoiding exceeding 4 layers. This is because using more than 4 layers can cause the lower grains to retain more moisture than the upper ones, which could result in uneven growth. In other words, the lower grains could develop faster than the upper ones.

Below are other examples of alternatives for carrying out the sowing process.

Note: You can experiment with different quantities of grains to determine which yields the best results in your case.

6-First phase/darkness:

This step is part of the first phase of production, from the first day in the trays until the day the seedling begins to sprout its first leaf, which may vary depending on the type of grain and the care given to the crop. Normally, this first phase lasts from day 1 to day 6. The way to know when this first phase ends is when the grain starts to sprout its first leaf.

During this initial stage, there are two fundamental objectives to ensure success in the production of hydroponic fodder. The first objective is to keep the grains hydrated at all times, especially during the first days of their life when they are more fragile. As they grow, they become more resistant, but in their early days, it is crucial to prevent them from drying out or losing their hydration. The second objective is to keep the grains in darkness, simulating their natural environment underground. This helps the grain develop its root and stem.

To achieve these two objectives, there are two techniques that can be implemented in both small and large-scale productions of hydroponic fodder:

1. Using black plastic:
Covering the area where the fodder is produced with black plastic, as shown in the image. The use of black plastic helps keep the grains in darkness, promoting root and stem growth, while also helping to keep the grains hydrated. It is important to remember that grains need to breathe, so some openings should be left in the plastic to allow for airflow.

2. Using a cloth or fabric: Cover the trays or containers with a cloth or fabric, as shown in the image. This also helps to keep the grains in darkness to promote root and stem growth, while also keeping the

grains hydrated. Watering is done over the cloth or fabric, allowing it to soak and thus maintain the grains' hydration. It is important to note that using a cloth or fabric is recommended due to the small holes that allow minimal airflow for the seedlings to breathe.

Note: When using commercial climate-controlled containers, these parameters are controlled 24 hours a day, so it would not be necessary to implement the two techniques mentioned above.

7-Second phase/illumination:

This step is part of the second stage of production, which spans from the day the grains start sprouting their first leaf until harvest day. The duration of this phase can vary depending on the type of grain and the care the crop receives. Typically, this second stage ranges from day 6 or 7 until day 12, which is the maximum recommended for harvesting. We can identify the start of this second stage when the grains begin to show their first leaf.

During this stage, the primary goal is to provide seedlings with some light to initiate their photosynthesis process. Think of plants as chefs working with sunlight, water, and air. During the day, when the sun shines, plant leaves capture that light and convert it into energy. Then, they take in water and air from the environment. With all these magical ingredients, plants perform a special recipe inside their leaves called photosynthesis, creating their own food called glucose. Additionally, as an extra gift, plants also produce fresh air (oxygen) for us all to breathe.

In this stage, the seedlings already have their roots and their first leaves, so they are stronger and can withstand longer periods without watering. Therefore, it is no longer necessary to keep them in darkness.

To reach this objective, artificial light or sunlight, as demonstrated in the next chapter, can be utilized. However, it's crucial to note that direct sunlight isn't advisable for hydroponic fodder production. At this stage, plants are only germinated and haven't reached their reproductive phase.

In summary, when producing hydroponic fodder using sunlight, our goal is to provide just enough light to initiate the process of photosynthesis. As can be seen in these images, direct sunlight is not present in any of them.

8-Harvest:

The final step is harvesting our hydroponic fodder, which can be implemented in animal diets at different times, ranging from the first 3 to 6 days to more than 14 days after planting. Here, the important thing is to analyze the reason for producing this fodder. It is recommended to harvest when the nutritional levels are optimal, which can be identified in two ways: when it reaches a height between 8-10 inches or between day 10 and day 12, without exceeding 12 days or 10 inches, as after that time, the nutritional levels begin to decrease. When all factors are controlled 100% (temperature, humidity, luminosity, and aeration), a height of 8-10 inches can be reached within approximately 6-10 days. Conversely, if these factors are not controlled, it is likely that this height will not be reached, and the fodder should be harvested no later than day 12, regardless of the weight and height of the fodder.

On the other hand, if you need the fodder to combat drought or scarcity, it is common to harvest after 12 days, even if the nutritional levels are not at their maximum. In these cases, the main goal is to ensure the survival of the animals during extreme periods, prioritizing the production of the highest final weight of our hydroponic fodder. Although the nutritional values may not reach their maximum, this fodder will still be nutritious. To precisely know the nutritional values, it is essential to perform an analysis.

For chickens, it is common to harvest between day 3-6, as they prefer germinated grains. It is recommended to provide them with hydroponic fodder of wheat with just six days of germination, as it improves digestion compared to only feeding them grains.

It is important to note that nutritional values may vary depending on the type and variety of grains used, as well as the care provided to the crop.

Below, I share the link to the video where you can observe these production steps in more detail and appreciate the final results of our hydroponic fodder.

"8 steps to produce Hydroponic Fodder"
Youtube: https://www.youtube.com/watch?v=YN6pauy-i6g
Facebook: https://fb.watch/qSHTHiwzHU/

Factors Influencing Production

As we have previously highlighted, the successful production of hydroponic fodder is based on four fundamental factors: Temperature, Humidity, Light, and Aeration. In summary, the quality of your results will be directly related to the ability to maintain these factors in optimal conditions. Each one plays a crucial role in the growth and development of hydroponic fodder, and their proper management is essential for obtaining effective and nutritious yields.

Temperature:

Temperature plays a crucial role in plant germination, influencing water absorption and evaporation. Significant fluctuations in this factor can impact crop yield. Effectively managing temperature is vital for hydroponic fodder production. Generally, the optimal temperature range for hydroponic fodder production is between 64°F and 78°F.

However, different grain species have varying temperature requirements for germination. For instance, grains such as oats, barley, and wheat thrive in cooler temperatures, ranging from 64°F to 70°F. On the other hand, grains like corn prefer warmer climates, with the ideal temperature for corn hydroponic fodder falling between 77°F and 82°F (Martínez, E. 2001; personal communication).

As the minimum germination temperature increases, proper drainage of the trays becomes crucial to prevent excess moisture and the proliferation of fungal diseases. Constant monitoring and quick response to abnormal situations are essential, as fungal attacks can be devastating to production in a matter of hours.

For effective management, the installation of a maximum and minimum hygrometer in production areas is recommended. This facilitates daily temperature control and early detection of potential problems resulting from variations outside the optimal range.

Establishing the hydroponic fodder production system in environments isolated from external climate changes significantly contributes to optimizing production.

Humidity:

Water, an essential element in the life of plants, is supplied through irrigation, making it a factor of vital importance. Careful management of humidity within the production area is fundamental.

However, a relative humidity above 80%, without adequate ventilation, can lead to phytosanitary problems, especially difficult-to-control fungal diseases. Conversely, too much ventilation can lead to crop dehydration, significantly affecting production. Therefore, finding a balance between the percentage of relative humidity and the optimal temperature is key to success in hydroponic fodder production.

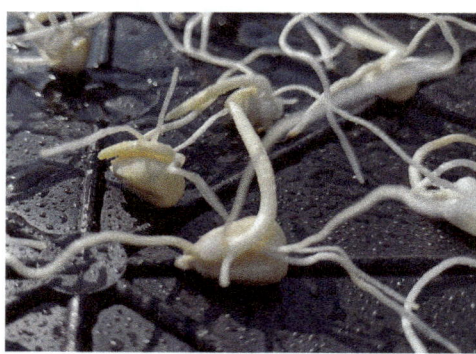

(A humidity range between 60% - 80% is ideal).

Humidity plays a crucial role in the cultivation of hydroponic fodder, as it affects various aspects of the plant growth process. Here are some ways in which humidity can influence hydroponic fodder:

1. Seed Germination: Humidity is essential to initiate the seed germination process. A humid environment helps activate the seeds and facilitates water absorption, allowing them to germinate and develop.

2. Seedling Development: During the early stages of growth, seedlings require a humid environment to develop properly. Air humidity influences water absorption by the roots, which affects the growth and health of the seedlings.

3. Evaporation and Transpiration: Ambient humidity affects the rate of water evaporation from the cultivation system and plant transpiration.

4. Disease Control: Excessive humidity can create a conducive environment for the growth of fungi and bacteria, increasing the risk of diseases in the crop. On the other hand, humidity levels that are too low can increase the incidence of water stress in plants.

In summary, appropriate humidity is crucial for the success of hydroponic fodder cultivation. A proper balance of humidity is essential to ensure healthy plant growth and avoid issues such as diseases and water stress.

Luminosity and Artificial Light:

It is important to remember that high levels of brightness are not required, as hydroponic fodder simply involves the germination of grains and does not reach the reproductive stage of plants, as is the case with corn.

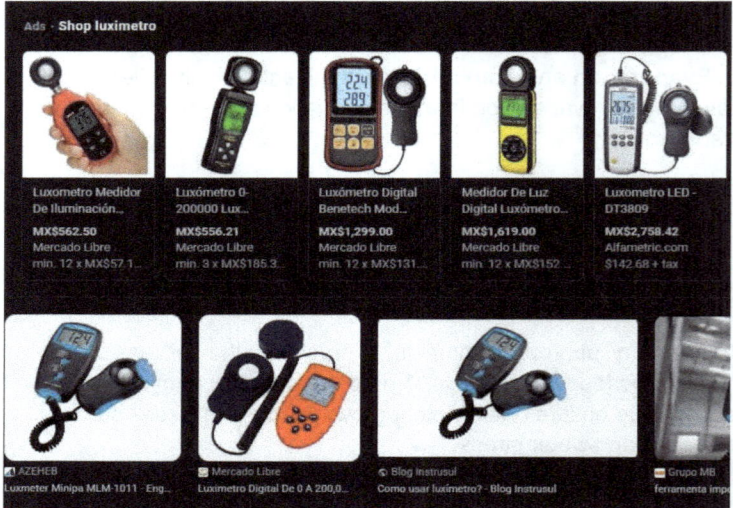

Brightness is measured with a lux meter, which records lux levels.

Although direct sunlight is not recommended, it is crucial to provide some brightness to support the plants' photosynthesis processes. Without light in the premises for hydroponic fodder, the green cells of the leaves cannot carry out photosynthesis, and therefore, there will be no biomass production.

The optimal brightness ranges between 30,000 Lux - 50,000 Lux, and it has been produced with a minimum brightness of 2,500 Lux - 3,000 Lux. When brightness exceeds 50,000 Lux, plants experience stress, and the roots may begin to dry out. For this reason, direct sunlight is not recommended for hydroponic fodder production.

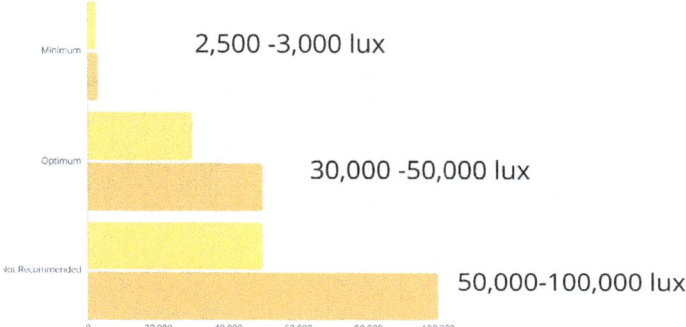

2,500 -3,000 lux

30,000 -50,000 lux

50,000-100,000 lux

During the seed germination stage in the hydroponic fodder production cycle, direct exposure to light is avoided during the initial days to promote sprouting and root development. In enclosed facilities, towards the end of the production process, trays are exposed to light to achieve the characteristic intense green color of the fodder and complete its optimal nutritional richness. In cases of exclusive production in enclosed spaces without natural light, the installation of artificial lighting through well-distributed fluorescent tubes should be considered.

In the pictures below, you'll notice that the light shining on the plants is indirect. There's no direct sunlight in any of these images. Instead, the plants are only receiving light that has bounced off surfaces nearby.

36

As we can see below, artificial light is being used. This type of lighting comes in various price ranges, and there are several options available in the market.

The main difference between household bulbs (standard light bulbs) and bulbs designed to provide light to plants (growth lights or grow lamps) lies in the spectrum of light they emit.

In summary, bulbs designed for plants are specialized to provide the necessary light for the process of photosynthesis and the healthy growth of plants, while household bulbs are more generic and do not always offer the appropriate light spectrum for optimal plant development.

Below in this image are different options of artificial lights specially designed for plants. Explore some of the alternatives available in the market to illuminate and promote the healthy growth of your crops.

Aeration or Ventilation:

Ventilation is crucial in the production of hydroponic fodder for several reasons:

1. Oxygen for Roots:
Adequate ventilation ensures that plant roots receive enough oxygen. Oxygen is essential for root respiration and overall plant function. Insufficient ventilation can deprive roots of necessary oxygen, affecting plant growth.

2. Prevention of Fungal Problems:
Good ventilation helps reduce moisture in the growing area. Without sufficient air circulation, fungi and bacteria can thrive in overly humid conditions. Proper ventilation prevents fungal diseases that could harm hydroponic fodder.

3. Temperature Control:
Ventilation is essential for maintaining the correct temperature in the growing area. Excessive heat can negatively impact plant growth. Proper ventilation disperses heat and maintains optimal conditions.

4. Avoidance of Harmful Gases:
In closed systems, photosynthesis can produce carbon dioxide and other gases. Adequate ventilation prevents the buildup of these gases, ensuring a continuous supply of fresh air for plants.

5. Air Quality Improvement:
Ventilation contributes to air renewal, enhancing overall air quality in the growing area. This is important for providing plants with the proper gas mixture and facilitating nutrient absorption.

6. Optimal Development:
Well-ventilated environments foster optimal plant development by providing ideal conditions for processes like photosynthesis and transpiration.

In summary, ventilation is vital for the successful growth of hydroponic fodder. We need fresh air and the removal of stale air wherever we cultivate fodder. This helps prevent issues like excess water and diseases. Therefore, it's essential to ensure good airflow to keep plants happy and growing healthily.

Systems and Shelving

In this book, we will learn how to adjust the conditions to produce two types of Hydroponic Fodder: one in vertical tower trays and the other in horizontal matting. For a Hydroponic Fodder system to work well, it is crucial to pay attention to all the details and take care of each aspect of the technique.

Vertical System

This method involves creating a shelving system to produce Hydroponic Fodder. A variety of materials can be used to build these shelves, including aluminum, PVC, wood, and other metals such as steel, sheet metal, or construction rods. Plastic trays are used for planting, ensuring efficient use of space in areas where traditional soil cultivation is challenging. Additionally, these shelves can also be purchased pre-made, facilitating their implementation in various environments.

A vertical tray system for Hydroponic Fodder is a structure where we cultivate fodder vertically. Picture trays stacked on top of each other, like a tower or a pyramid. In this system, we plant the seeds in the trays and water them. The plants grow in a controlled environment and, after a while, they are ready to be harvested.

The advantage of this method is that it allows us to cultivate a lot of fodder in a small space, ideal for places with limited space. Additionally, the vertical arrangement makes it easier to care for and manage the plants.

(It is important to note that aluminum and PVC are the most common materials for these shelves, as they are lightweight and durable. In comparison, wood tends to rot over time, and metal can rust).

When setting up shelves for Hydroponic Fodder, regardless of the material you choose, there are two important things you must ensure for successful production. Here they are explained in simple terms:

1. Inclination: The shelves need to have a slope, like a small incline, so that excess water can drain out when you water the plants. This prevents too much water from accumulating. **The most common inclination is around 12° to 13°, which means approximately 2 to 3 inches of incline. This way, when we water, the excess water can drain out properly, preventing it from accumulating where we don't want it.**

2. Space between Levels: You also need to leave space between each level. This helps distribute the water evenly and allows the hydroponic fodder to grow well. Each level needs its own space.

Plants need about 8-10 inches to grow well. We also need an extra 2 inches for watering them evenly. In total, it's most common to have about 12 inches of space between each level. This way, the plants can grow happily and receive the water they need.

The inclination can always be adjusted depending on the size of the trays; some trays may be longer than others. The primary goal is to prevent water accumulation, which in turn helps avoid excess humidity. Adjusting the inclination ensures that water drains properly, maintaining optimal conditions for plant growth.

Characteristics of Hydroponic Fodder Trays

In the market, there are various trays designed for hydroponic fodder cultivation, and although they may vary, they share characteristics that are beneficial for plants. Below, I describe some of these features:

1. Black Color: These trays are usually black, which simulates the underground conditions where plant roots normally grow. This aspect contributes to a visual environment that favors plant development."

2. Drainage Holes: Equipped with holes, these trays allow efficient drainage of excess water, preventing harmful accumulation for the roots. This feature is crucial for maintaining proper water balance.

3. Flat Surface: The flat shape of the trays facilitates uniform distribution of water across the entire surface, ensuring that all plants receive the necessary amount for optimal growth.

4. Sturdy: The rigidity of these trays provides durability and resistance, preventing deformations that could compromise their functionality. This strength facilitates handling, washing, and care, allowing for prolonged use over time.

How to Build a Homemade System

As previously mentioned, you have the flexibility to construct your system using various materials, and there are multiple tray sizes offered in the market. Consequently, there are no exact measurements that I can provide. However, if you opt to create your system, the following steps are common to all systems:

1. Choose the necessary materials:

- Choose a material for the shelf, such as metal, aluminum, wood, PVC, etc.

- Acquire the germination trays or seedling trays before making any cuts to ensure compatibility.

2. Construction of the shelves:

- Make the necessary cuts according to the chosen material.

- Ensure that the height between each level is suitable for plant growth.

- Ensure that the trays are tilted for efficient drainage of excess water.

Irrigation System

Sprinkler and mist irrigation is quite common in hydroponic fodder production due to its lower cost. Regardless of whether you use sprinkler or mist irrigation, this irrigation system consists of three key components:

1. Hoses and Sprinklers: These components act as tubes and sprinklers that evenly distribute water over the plants to promote their growth.

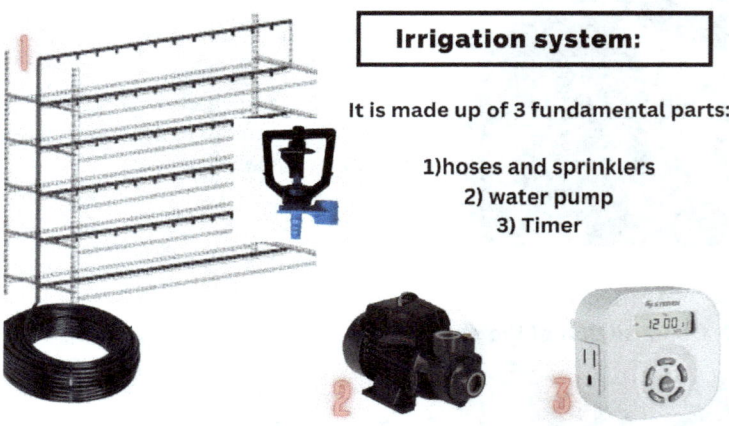

Irrigation system:

It is made up of 3 fundamental parts:

1) hoses and sprinklers
2) water pump
3) Timer

2. Water Pump: The water pump functions as the engine that drives and distributes water throughout the irrigation system.

3. Timer: This device is essential for programming and controlling the timing of irrigation cycles. This ensures that plants receive the appropriate amount of water at the right times.

In this image, we can see some of the most common sprinklers used for mist irrigation.

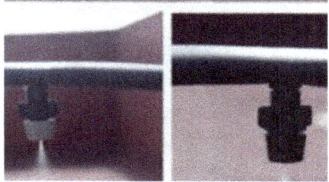

It is worth noting that you can also use PVC piping for the irrigation system, as shown in the image.

It is also important to note that there are various types of sprinklers available, and the main difference lies in the amount of water each one uses. In the market, you can find sprinklers ranging from 7 gallons per hour to over 26 gallons per hour.

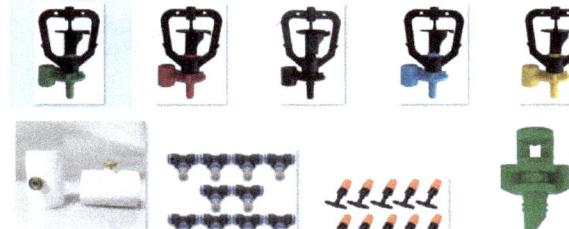

46

The amount of water they use is influenced by the water pump pressure. For example, green sprinklers use between 16 and 19 gallons per hour. Operating at 3.3 lbs of pressure, they have a watering radius of 10 feet, and operating at 5.5 lbs of pressure, the watering radius is 12 feet. For this reason, it will be important to conduct tests to determine the specific range of the sprinklers and water pump you choose.

Regardless of the sprinklers you choose, it is common to use at least 2 of them per level to ensure uniform irrigation, especially in stands of 5 to 7 feet in length. However, this quantity may vary depending on the pressure and specific type of sprinkler you are using.

Drip Irrigation

It is also possible to adapt these systems to drip irrigation, although it is not very common due to the excessive use of water. However, an

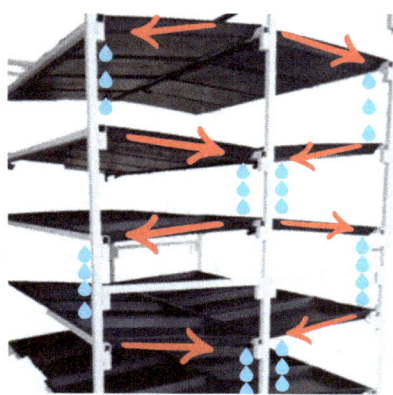

advantage of this method is that electricity is not needed for its operation. In vertical systems, the trays or shelves are tilted so that irrigation starts in the upper trays, allowing water to move downward, progressively watering the lower trays.

Here you can see other examples of drip irrigation.

Horizontal System:

This method stands out for its economic efficiency, making use of materials readily available in the local market without the need for costly structures. Despite not requiring these advanced installations, it is advisable to use a shade net cover to protect the crop from intense sunlight and prevent interference from unwanted insects or animals, such as rodents or squirrels, commonly found in outdoor environments.

The versatility of this system is demonstrated by constructing it with various materials, as illustrated in the attached image. The key to this setup is applying a ground slope of 10-12%, ensuring effective drainage for both rainwater and scheduled irrigation, thus avoiding harmful accumulations. It is essential to highlight that, regardless of the system's orientation, whether vertical or horizontal, it is always crucial to have a proper slope to facilitate the drainage of excess water, as waterlogging is detrimental to the plant roots' development.

This type of system can also be integrated with a vertical system, as shown in the following image, in this image we can see 10 feet-long wood arranged vertically to maximize space. When using wood, it is essential to use black plastic, as the wood could deteriorate and cause fungal and bacterial problems in the hydroponic fodder.

It is also quite common to implement this type of system directly on concrete. When used directly on the ground, it is essential to use black plastic. It is important to note that the use of white or transparent plastic is not recommended because the goal is to simulate that the roots are underground. White white or transparent plastic allows light to pass through, which can cause stress to the roots, unlike black plastic, which promotes the growth and development of roots and stems.

Nutrient Solution

I'd like to start by mentioning that many studies confirm that it's possible to achieve positive outcomes by using only drinkable water or groundwater without needing to include nutrient solutions. Furthermore, it is relevant to mention that the seeds themselves contain all the necessary nutrients for germination and initial plant growth. It is important to note that Hydroponic Fodder is limited to being a sprout that develops over a period ranging from 7 to 14 days. The key lies in understanding that during this initial phase, plants do not require intensive nutrition, as the seed already provides the essential nutrients for its early development.

Although a nutrient solution may contribute to increasing the final weight and improving nutritional values, true success lies in providing optimal conditions of temperature, humidity, luminosity, and aeration.

It is crucial to highlight that the use of nutrient solution carries its risks, as while it provides beneficial nutrients, it can also promote the growth of fungi and bacteria.

For those starting hydroponic fodder production for the first time, it is recommended to begin with plain water. After achieving initial success, the gradual introduction of nutrient solution can be considered, always mindful of the potential risks associated with fungi and bacteria.

What type of water should I use?

The quality of irrigation water is essential for success in hydroponic fodder cultivation. The water should be potable, whether from a well, rain, or public supply. If the water is not potable, sanitary issues may arise.

When the water quality is not optimal, a detailed chemical analysis should be conducted. Based on the results, it is necessary to reformulate the nutrient solution and consider other treatments such as filtration, settling, sun exposure, acidification, or alkalization to ensure its quality.

What is a nutrient solution?

A nutrient solution (NS) consists of water with oxygen and all essential nutrients in ionic form, possibly with some organic compounds such as iron chelates and other micronutrients (Steiner, 1968). A genuine NS is one that contains the indicated chemical species, matching the corresponding chemical analysis (Steiner, 1961).

Chemical Species

$$H, H^+, H_2, H_2^+, H_3^+, H^-$$
$$He, He^+, Na, Na^+, e^-$$
$$C, C^+, CH, CH^+, CH_2, CH_2^+$$
$$CH_3, CH_3^+, CH_4^+, CH_4, CH_5^+$$
$$O, O^+, O_2, O_2^+, OH, OH^+, H_2O, H_2O^+, H_3O^+$$
$$CO, CO^+, HCO, HCO^+, H_2CO, H_2CO^+$$

The NS follows the laws of inorganic chemistry, with reactions that form complexes and precipitate ions, allowing them to be available to the plant roots (De Rijck and Schrevens, 1998).

Functions of Nutrients in Plants:

Below, we will explore the 13 essential nutrients for the growth and development of plants in general.

Macronutrients	Micronutrients
• Nitrogen (N)	• Zinc (Zn)
• Phosphorus (P)	• Iron (Fe)
• Potassium (K)	• Manganese (Mn)
• Sulfur (S)	• Copper (Cu)
• Magnesium (Mg)	• Chlorine (Cl)
• Calcium (Ca)	• Boron (B)
	• Molybdenum (Mo)

The nutrients that serve specific functions in plants are divided into the following three main groups:

1. Structural: These nutrients are part of organic compounds, such as amino acids and proteins (N), pectates (Ca) in the middle lamella of the cell wall, and (Mg) in the center of chlorophyll molecules.

2. Enzyme Constituents: These are elements, typically metals or transition metals (Cu, Fe, Mn, Mo, Zn, Ni), that are part of the prosthetic group of enzymes essential for their functions.

3. Enzymatic Activators: These are dissociated components of the protein fraction of enzymes, necessary for their functions.

Preparation of the Nutrient Solution

Keep in mind that if you lack experience in preparing a nutrient solution or if it's difficult to find these nutrients in your area, you can obtain pre-made nutrient solutions from a hydroponic store or online markets like Amazon. In this section, we will show you how to prepare a successful formula using two stock solutions: Concentrated Solution A and Concentrated Solution B. The first contains mainly macronutrients (N, P, K, S, Mg, Ca), and the second contains micronutrients (Zn, Fe, Mn, Cu, Cl, B, Mo).

First, it's essential to note that in hydroponics, the composition of the nutrient solution varies depending on the plant's growth stage. The nutritional requirements are not the same for a flowering plant as they are for one in fruit development. Although the plant always needs the 13 essential nutrients, the quantity varies according to the production phase.

It's relevant to highlight that during the initial germination stage, the nutrient requirements are lower compared to a plant in full development. This is important because hydroponic fodder, being simply a sprout, doesn't require a significant amount of nutrients at this stage.

As mentioned earlier, the seed already contains everything necessary for its initial growth. Although a nutrient solution can contribute to the final weight and nutritional value, the real success lies in providing optimal conditions of temperature, humidity, luminosity, and aeration.

These factors are fundamental for the healthy development of plants and must be carefully controlled to maximize the production and quality of hydroponic fodder.

The formula for calculating the nutrient solution can be expressed in grams per liter (g/l) or in parts per million (ppm) of the selected nutrients. Below, I provide you with information based on parts per million (ppm).

Over nearly 200 years of research and observation in the field, the following table has been compiled. While we cannot consider it exact in all cases, as the requirements of plants, like living beings, can vary, this table can serve as a guide to determine how many parts per million of nutrients most plants need.

ESSENTIAL ELEMENTS, MACRONUTRIENTS, MICRONUTRIENTS AND THEIR MOLECULAR WEIGHT

	Molecular weight	Maximum (ppm)	Optimum (ppm)	Minimum (ppm)
N	14.0067	300	200	47
P	30.973762	130	60	30
K	39.0983	600	400	50
Ca	40.078	400	250	50
Mg	24.3247	150	50	25
S	32.065	650	70	50
Fe	55.845	9	5	2
Mn	54.938049	1.6	0.8	0.5
B	10.811	2	0.5	0.25
Cu	63.536	0.1	0.05	0.005
Mo	95.94	1.6	0.8	0.2
Zn	65.409	0.75	0.5	0.05
Si	28.0855	0.05	0.01	0.005
C	12.0107	1500	600	250

Once you have a formula, which can be any of those in this last image (Maximum ppm, Optimum ppm, or Minimum ppm), the next step is to determine the quantities of fertilizers to be used. This will depend on the fertilizers available in your area and their commercial names.

The simplest way to calculate the exact amount of nutrients is by using an Excel spreadsheet, which is easy to find on the Internet. Just search for "nutrient solution calculator" or "Excel spreadsheet for nutrient solution," and upon entering the formula, the Excel sheet will indicate how many grams of each nutrient you need. There are also free applications available that you can download to your devices.

Below are some of these applications and websites that can be useful.

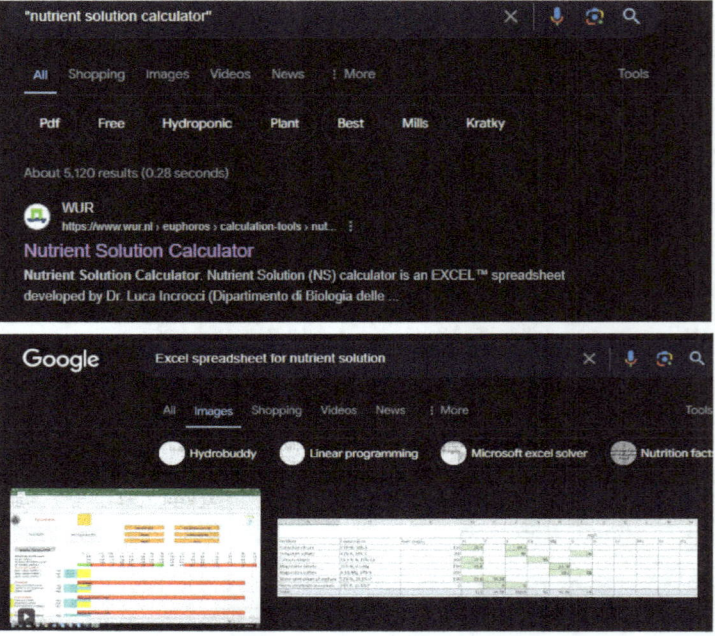

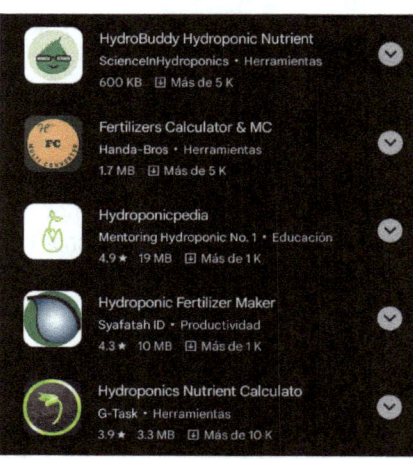

Free applications available for download on your devices.

57

Below, I share the nutrient solution that I have personally developed, specifically for my Hydroponic Fodder. Over several years, I have made modifications that have resulted in better yields. Additionally, I will provide the commercial names of the fertilizers and the exact amounts of each of them. This nutrient solution is designed to be used in 1000 liters (264 gallons) of water.

Nutrient Solution (PPM)

N-128	N-Nitrogen
P-27	P-Phosphorus
K-175	K-Potassium
Ca-100	Ca-Calcium
Mg-0.8	Mg-Magnesium
Fe-2	Fe-Iron
Zn-0.1	Zn-Zinc
Cu-.07	Cu-Copper
B-.3	B-Boro
Mo-.03	Mo-Molybdenum
S-26	S-Sulfate

In the first row, we have the required amounts of nutrients measured in parts per million (ppm). The second row shows the full names of these nutrients.

As we can observe, the nutrient quantities are quite low, even compared to the previously mentioned table. Below, I present the commercial names of the fertilizers and the quantities in grams of each of them. These fertilizers in my country can be found online. Similarly, on platforms like Amazon, you can even purchase them ready to use. We note that many of these fertilizers come in presentations of 25-50 kilograms (55-110 lbs), which is a considerable amount for our use. Therefore, at times, it may be more practical to acquire these solutions already prepared for use.

Grams per 1,000 liters of water

Solu potase-417 gm
Sulfo nit-8gm
Multimap-61gm
Calcinit-526gm
Sulmag-125gm
Queleto Fe6%-11gm
Fertiman-3gm
Boric Acid-1gm
Fertizinc-.2gm
Comet-.2gm
Sulfato de amonio-2gm

   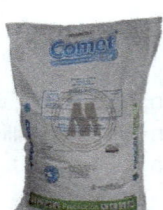

Understood, it's time to prepare the nutrient solution for our hydroponic fodder.

Preparing our nutrient solution is straightforward. The key here is to dissolve these fertilizers in 2 parts. In one bucket of water, we'll dissolve the Macronutrients, and in another bucket, we'll dissolve the micronutrients. Finally, we'll combine both solutions, Solution A and Solution B.

Solution A

Solu potasse- (417 grams)
Sulfo nit- (8 grams)
Multimap- (61 grams)
Calcinit- (526 grams)

Solution B

Sulmag- (125 grams)
Fertiman Soluble- (3 grams)
Boric Acid- (1 grams)
Fertizinc- (.2 grams)
Comet-(.2 grams)
Sulfato de amonio- (2 grams)
Quelato Fe6%- (11 grams)

Procedure: Solution A

In a plastic container, measure 2 gallons of water and add each of the elements previously weighed in the mentioned order. Begin constant stirring and add the second nutrient only when the first one has completely dissolved. Then, add the third one after the complete dissolution of the previous two. Once few residues of the applied fertilizers remain, top up with water until reaching 3 gallons and stir for an additional 10 minutes, ensuring there are no visible solid residues. This way, we obtain Concentrated Solution A. If using it for a container of less than 1000 liters (264 gallons), the remaining amount of solution should be packaged in a demijohn or glass container, labeled, and stored in a dark and cool place.

Procedure: Solution B

In a plastic container, measure 1 gallon of water and add each previously weighed element, respecting the order to ensure complete dissolution. Finally, incorporate the Iron Chelate (Quelato Fe6%). Proceed to dissolve the mixture for at least 10 minutes, ensuring no solid residues remain from any component. Then, top up the volume with water until reaching 2 gallons and stir for an additional 5 minutes. This way, we obtain Concentrated Solution B.

Observations of solution A and B

It is essential not to exceed the recommended amounts, as it could lead to crop poisoning. The water used for preparation should be ordinary tap water at normal temperature (68°F to 77° F), although distilled water is preferable if its cost is not high. When handling nutrients, whether concentrated, in preparation, or as a nutrient solution, plastic or glass materials should be used, avoiding metal or wooden stirrers.

In the preparation of the NUTRIENT SOLUTION applied to the crop, it is crucial to avoid mixing CONCENTRATED SOLUTION A with CONCENTRATED SOLUTION B without the presence of water. This practice would activate a large portion of the nutrients in both solutions, causing more harm than good to the crops. The proper mixing is achieved by adding each solution separately in water, first one and then the other.

pH (hydrogen potential) & EC (electrical conductivity)

When using a nutrient solution (NS), it's crucial to check and adjust the pH and EC levels. This is important because it ensures that the plant can properly use the nutrients. If the pH and EC aren't at the right levels, the plant won't be able to benefit from them. Therefore, applying the nutrient solution would be pointless. Below, we'll explain how to check and modify these two parameters to achieve the best results.

Control of pH (Hydrogen Potential)

pH (Hydrogen Potential) is a system that measures the concentration of hydrogen ions in a solution, being crucial for the optimal growth of plants. The nutrient solution used in Hydroponic Fodder consists mainly of water, composed of hydrogen ions (H+) and hydroxyl ions (OH), which combine to form water (H2O or HOH).

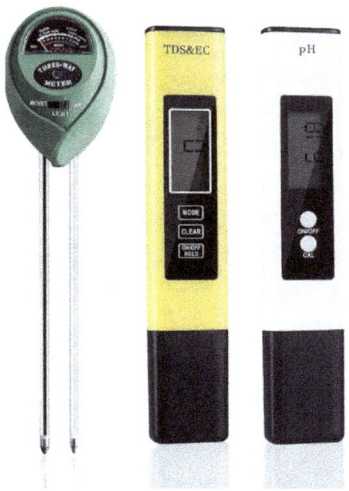

Therefore, it is essential to monitor these components using a pH meter.

The pH value of our nutrient solution can range from 5.2 to 7.5, **preferably maintaining it between 6.5 and 7**, with some exceptions, such as legumes, which can thrive with a pH close to 7.5. However, most seeds, especially cereals used in Hydroponic Fodder, do not perform efficiently above a pH value of 7. It is important to note that plants absorb nutrients better when the pH is within optimal conditions. Maintaining the pH within these ranges is essential to promote an environment conducive to growth.

In the following image, we'll explore how plants take in nutrients depending on the pH level, and why keeping the pH at an optimal range is crucial for plants to utilize these nutrients effectively.

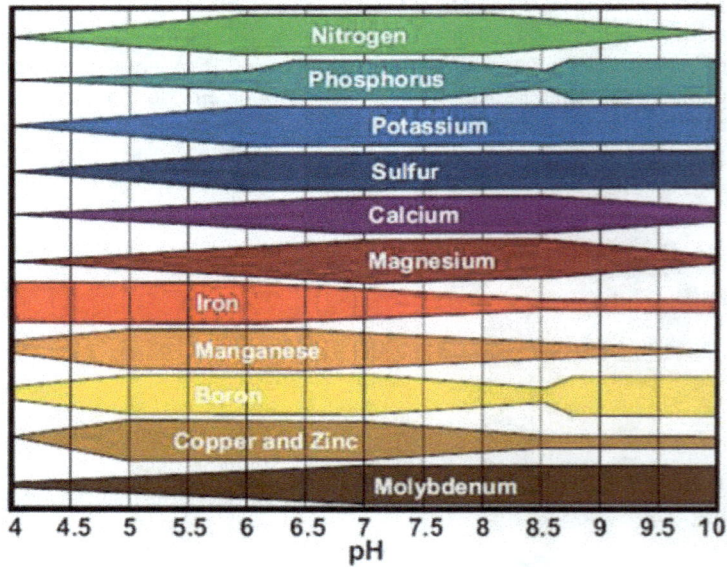

pH Observation:
To adjust the pH in a nutrient solution, specific substances can be used. Here are some common options:

To increase the pH (make the solution more alkaline):

1. Potassium Hydroxide (KOH): It is a strong base used to raise the pH.

2. Sodium Carbonate (Na2CO3): Also known as soda ash, it can elevate the pH of the solution.

To decrease the pH (make the solution more acidic):

1. Phosphoric Acid (H3PO4): It is a commonly used acid to lower the pH.

2. Nitric Acid (HNO3): Another acidic option to decrease the pH in nutrient solutions.

It's important to measure the pH with a reliable pH meter and adjust gradually, avoiding sudden changes. Careful pH control is essential to ensure that plants can effectively absorb nutrients.

Control of EC (Electrical Conductivity)

The electrical conductivity (EC) in hydroponics refers to the water's ability to conduct electrical current. In the context of hydroponics and hydroponic fodder, EC is an important measure to monitor the concentration of dissolved salts in the nutrient solution used for soilless plant cultivation.

An appropriate EC is crucial for healthy plant growth. If EC is too low, it may indicate a lack of essential nutrients for the plants. Conversely, if EC is too high, it can lead to toxicity, which can also negatively affect plant growth.

In the specific case of hydroponic fodder, which is a method of cultivating fodder for animal feed, the EC of the nutrient solution is essential to ensure optimal plant growth and, therefore, adequate nutritional quality of the fodder produced.

In summary, EC in hydroponics and hydroponic fodder is an important measure to ensure a proper balance of nutrients and to avoid salt toxicity problems in plants.

The recommended EC for hydroponic fodder cultivation is usually in the range of **1.5 to 2.5 mS/cm (millisiemens per centimeter)**. This range provides an adequate amount of dissolved salts for healthy plant growth without causing salt excess toxicity.

To control EC in hydroponic fodder cultivation, the following steps can be followed:

1. Use of EC meters: Specific EC meters can be used to obtain accurate readings of the water's electrical conductivity in the nutrient solution. These devices are designed to be immersed directly in the solution to take measurements.

2. Adjustment of the nutrient solution: If the measured EC is outside the desired range, adjustments can be made to the nutrient solution. This may involve adding more water to reduce the salt concentration or adding more nutrients to increase it, as necessary.

3. Monitoring of nutrient levels: In addition to controlling EC, it is important to monitor the individual nutrient levels in the nutrient solution,

such as nitrogen, phosphorus, potassium, and micronutrients. This will help ensure that plants receive all the necessary nutrients for healthy growth.

4. Records and constant adjustments: Keeping records of EC readings and any adjustments made to the nutrient solution is essential to maintain accurate and consistent EC control over time. This allows for quick adjustments in case of fluctuations.

Properly controlling EC in hydroponic fodder cultivation is essential to ensure optimal plant growth and adequate nutritional quality of the fodder produced.

Application of the Nutrient Solution

It is advisable to start using the Nutrient Solution 2-3 days after starting the procedure and to suspend its use 2-3 days before harvesting for two fundamental reasons, both of great importance.

1. In the initial days, the grain does not require nutrition. As mentioned earlier, since the seed contains all the necessary nutrients for germination and initial plant growth in its early days of life, it is not necessary to apply the nutrient solution in this phase.

2. In the final days, it is advisable to use only water, as the nutrient solution can generate unpleasant odors that discourage animals from consuming the Hydroponic Fodder. Furthermore, this solution contains chemical fertilizers that, in large quantities, can cause health problems for farm animals. Therefore, the recommendation is to use only water in the days leading up to harvest to eliminate any residue of these chemical fertilizers.

As shown in the table below, in a 12-day cycle, the nutrient solution is only applied for 6-8 days. In the first row, it can be observed that the nutrient solution is applied from day 3 to day 10, while in the second row, it can be seen that the nutrient solution is applied from day 4 to day 9. These are just examples of how this nutrient solution can be applied to ensure it is not applied when not needed and to prevent these chemical fertilizers from negatively affecting the health of our animals.

1	2	3	4	5	6	7	8	9	10	11	12
1	2	3	4	5	6	7	8	9	10	11	12

An interesting alternative in hydroponic fodder production is the use of worm humus instead of conventional nutrient solutions. Worm humus, also known as vermicompost or worm compost, can be an excellent source of nutrients for plants.

The liquid worm humus can be directly incorporated into the water used for irrigating the plants. Worm humus provides a variety of essential nutrients and beneficial microorganisms that can enhance the health and growth of hydroponic fodder plants.

It's important to note that the doses of worm humus may vary depending on the manufacturer, as not all worm humus contains the same amounts of nutrients. Therefore, it's advisable to consult with the supplying company to determine the recommended amount of worm humus per gallon of water to be used in the irrigation system. This will ensure proper and optimal application of this organic fertilizer.

Frequency for irrigating our hydroponic fodder each day?

Ensuring that irrigation is done at the right intervals is essential to meet the needs of the crop, its density, and specific weather conditions. It is of utmost importance to choose suitable sprinklers and precisely control the irrigation duration to prevent the occurrence of fungus in the fodder. Adapting the intervals and amount of water applied is crucial to avoid excess that could negatively affect the development of Hydroponic Fodder and potentially impact the health of the animals consuming it. The irrigation frequency is adjusted according to the climate of the production area. In cold and humid areas, irrigation is generally done between 3 and 5 times a day, while in hot and dry places, the frequency can be increased to 8 or 9 times a day, with short periods of 30 seconds to 1 minute per irrigation.

The main goal of irrigation is to ensure constant hydration of the seeds or grains, especially during their early days of growth. Given the vulnerability of plants in this initial germination phase, maintaining an adequate level of hydration is essential for their development. As plants grow, their resistance increases, allowing them to tolerate longer periods without the need for irrigation. Adjusting the intervals and duration is essential to ensure optimal growth and development of Hydroponic Fodder.

Steps to Calculate Hydroponic Fodder Irrigation Needs

Determining the irrigation intervals in your area is a crucial exercise to optimize the growth of your crop. From my experience, I've found that the following technique is easy to apply and, at the same time, provides excellent results to ensure the proper hydration of the

grains. This technique is very simple. All you need is a thin fabric or cloth, as shown in this image.

The first step is to determine whether we need one or two different irrigation intervals. In other words, we must decide if we require irrigation during the day and also during the night. In situations where temperature, humidity, light, and ventilation are constantly controlled 100% throughout the 24 hours, we will only need one irrigation interval set at specific times during the day.

On the other hand, if we are producing in an outdoor environment, in a greenhouse, or in an enclosed space where we cannot maintain control over temperature, humidity, light, and ventilation throughout the 24 hours of the day, it will be necessary to establish two different irrigation intervals: one during the day and one during the night. This is because the conditions will vary significantly between day and night.

If it's necessary to implement two irrigation frequencies, we should divide our daytime and nighttime hours. To do this, simply ask yourself at what time the sun rises in the morning and at what time it sets in the evening. For example, here in our area, the sun rises at 8 am and sets at 8 pm.

This indicates that our irrigation during the day will be done from 8 am to 8 pm, while irrigation during the night would be conducted from 8 pm to 8 am. 12 hours during the day and 12 hours at night.

Now, the next question is: how often will we schedule these irrigation sessions?

We will place the fabric in the location where we plan to produce Hydroponic Fodder, preferably on one of the germination trays. Then, we will apply irrigation over the fabric and simply wait for it to dry. By doing this, we will determine how long it takes for the fabric to dry. In our case, as seen in this image, it takes 2 hours and 20 minutes to dry. This indicates that we need to apply irrigation approximately every 2 hours to ensure adequate hydration of the seedlings. It is important to note that this example is implemented in an outdoor system with low humidity, so the fabric dries faster compared to an indoor space or a greenhouse where humidity could be higher.

This exercise indicates that we will need to apply irrigation every 2 hours starting at 8 am and ending at 8 pm. In conclusion, this would be a total of 7 irrigation sessions during the 12-hour day. In the following image, we can observe the timing of each irrigation.

The next step would be to determine the intervals for nighttime irrigation, from 8 in the evening to 8 in the morning. Usually, during the night, 1 or 2 irrigation are done, or sometimes they may not be necessary, depending on the nighttime climate. To set the schedules for these irrigation, we will repeat the exercise with the cloth. We will leave the cloth wet overnight, and in the morning before the first irrigation sessions, we will check if the cloth or fabric is still moist. If it

remains wet, it indicates that we do not need to water at night. But if the cloth is completely dry in the morning, it indicates that we need to apply at least 1 irrigation during the night. In our case, since our nighttime schedule is 12 hours, from 8 pm to 8 am, we apply irrigation at 2 am, which would be halfway through this schedule.

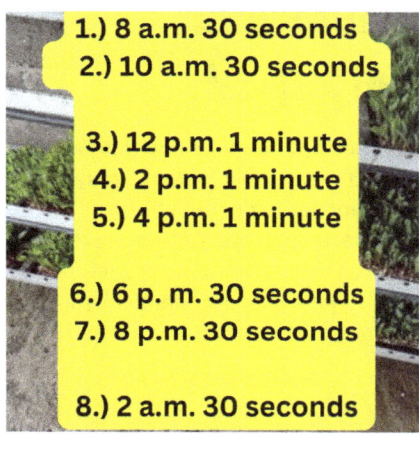

1.) 8 a.m. 30 seconds
2.) 10 a.m. 30 seconds
3.) 12 p.m. 1 minute
4.) 2 p.m. 1 minute
5.) 4 p.m. 1 minute
6.) 6 p.m. 30 seconds
7.) 8 p.m. 30 seconds
8.) 2 a.m. 30 seconds

In summary, we will need a total of 8 irrigation sessions, as shown in the attached image. The irrigation time will vary between 30 seconds and 1 minute, and this duration can be adjusted according to the specific needs of our hydroponic fodder and the climatic conditions of our area. Usually, 1-minute irrigation sessions are preferred during the hottest hours of the day, while 30-second irrigation sessions are chosen during cooler temperatures or when temperatures start to drop. This happens in the mornings when the sun rises and the day is still cool, as well as in the evenings when temperatures begin to decrease.

However, applying all irrigation sessions with a duration of 1 minute will not affect the growth of the crop; this time difference is primarily implemented to minimize water consumption to the necessary minimum. It is important to remember that the germination trays have holes and a specific slope to allow excess water to drain away.

Below, I share with you the link to one of our videos on YouTube and Facebook, where you can see how we carry out this exercise in our hydroponic fodder cultivation. Additionally, you will be able to see the final results obtained through the application of these 8 irrigation sessions.

Full video: "Sprinkler Irrigation Application in Hydroponic Fodder Cultivation | Irrigation Strategies"
Youtube: https://www.youtube.com/watch?v=FHmCFmXropA&t=1s
Facebook: https://fb.watch/qSEme6b7_s/

Use of Hydroponic Fodder for Different Species of Animals

There are various ways to incorporate hydroponic fodder into animal diets. It's important to remember that hydroponic fodder is only a part of animals' nutrition and it's not recommended to feed them exclusively with this type of fodder because it has very low levels of dry matter. All animal species have different nutritional requirements, and it would be important to consult with a professional to ensure a balanced diet. Below are the recommended quantities provided by the FAO.

The FAO (Food and Agriculture Organization) is a specialized agency of the United Nations dedicated to eradicating hunger, improving nutrition, and promoting sustainable agriculture worldwide. Its main goal is to achieve food security for all and ensure that people have access to enough quality food to lead a healthy life. The FAO works in collaboration with governments, international organizations, and civil society to develop policies, share knowledge, and provide technical assistance to countries on issues related to agriculture, food, and nutrition.

Recommended doses of Hydroponic Fodder (HF) by animal species (FAO)

	Animal species	Kg HF/ 100kg live weight	Observations
Quantities per day	Dairy Cattle	1 a 2	Supplement with straw, barley and other fibers.
	Dry Cow	0,5	Supplement with fiber. con fibras.
	Beef Cattle	0 a 2	Supplement with straw, barley and other fibers.
	Pigs	2	Supplement with concentrated feed.
	Poultry	25kg of HF/ 100 kg of dry food	Improves the conversion factor.
	Horses	1	Supplement with fiber and concentrated feed.
	Sheep/Goats	1 a 2	Add Fiber.
	Rabbits	0,2 a 2	Supplement with fiber and concentrated.

Source FAO (2001): food and agriculture organization

Below, I present a more detailed example of the recommended doses of hydroponic fodder per animal species according to FAO (2017). It is important to note that these quantities may vary depending on the type and variety of grain used, as well as the care provided to the crop. To establish an exact diet, it is recommended to conduct an analysis to accurately determine the nutritional values provided by our hydroponic fodder.

Dairy Cattle:
- Low Production: 33 lbs/day
- Medium Production: 44 lbs/day
- High Production: 62 lbs/day

Beef Cattle:
- Starter: 29 lbs/day
- Fattening: 37 lbs/day

Horses:
- Foals: 8.8 lbs/day
- Colts: 17.6 lbs/day
- Fillies: 8.8 lbs/day
- Empty Mares: 17.6 lbs/day
- Gestation: 8.8 lbs/day
- Stabled Horses: 15.4 lbs/day

Goats:
- Goats: 3.3 lbs/day
- Lactation: 5.5 lbs/day
- Dairy: 7.7 lbs/day
- Meat: 4.4 lbs/day

Sheep:
- Gestation Ewes (110 lbs): 5.5 lbs/day
- Lactation (1 lamb): 7.7 lbs/day
- Lactation (2 lambs): 8.8 lbs/day
- Meat: 6.6 lbs/day
- Lamb: 2.2 lbs/day
- Ram: 5.5 lbs/day

Pigs:
- Breeding Boars: 8.8 lbs/day
- Lactating Sows: 4.4 lbs/day
- Gestating Sows: 6.6 lbs/day
- Meat: 4.4 lbs/day

Rabbits:
- Gestation: 14.17 oz/day
- Lactation (6 kits): 19.05 oz/day
- Meat (30 days): 4.23 oz/day
- Meat (50 days): 6.35 oz/day
- Meat (70 days): 8.82 oz/day
- Meat (100 days): 12.69 oz/day

Chickens:
It is recommended to provide 55 lbs of Hydroponic Fodder per 220 lbs of dry feed. It is especially beneficial to supply wheat Hydroponic fodder with just six days of germination, as it improves digestion compared to exclusive grain feeding.

Bibliography

De Rijck y Schrevens, 1998.
Adams, 1994; Rincón, 1997.
Steiner, 1961.
La huerta hidropónica popular – FAO
Martínez, E. 2001; comunicación personal
Manual Técnico Forraje Verde Hidropónico – 2 Edición
https://cdigital.uv.mx/bitstream/handle/1944/50099/TeobaCruzMarco.pdf?sequence=1&isAllowed=y
https://www.fao.org/fao-stories/article/es/c/1375101/
https://www.fao.org/3/ah472s/ah472s00.pdf
https://ri.ujat.mx/bitstream/20.500.12107/3459/1/TESIS_LUIS_GUSTAVO_BALAM_LOPEZ.pdf
https://www.calameo.com/read/00527288440fda49deefe

www.ingramcontent.com/pod-product-compliance
Lightning Source LLC
Chambersburg PA
CBHW050236230526
45470CB00005B/1982